U.S. SEISMIC ZONES

THE CASCADE RANGE

GREG ROZA

PowerKiDS press

Published in 2026 by The Rosen Publishing Group, Inc.
2544 Clinton Street, Buffalo, NY 14224

First Edition

Editor: Greg Roza
Book Design: Michael Flynn

Photo Credits: Cover Scarola Photography/Shutterstock.com; (series background) Picksell/Shutterstock.com; p. 4 Bill Perry/Shutterstock.com; p. 5 zelvan/Shutterstock.com; p. 7 Bill Perry/Shutterstock.com; p. 8 https://commons.wikimedia.org/wiki/File:Alfred_Wegener_ca.1924-30.jpg; p. 9 Naeblys/Shutterstock.com; p. 11 Yarr65/Shutterstock.com; p. 13 Protasov AN/Shutterstock.com; p. 15 Michael Warwick/Shutterstock.com; p. 16 Deni_Sugandi/Shutterstock.com; p. 17 Rainer Lesniewski/Shutterstock.com; p. 19 Laima Swanson/Shutterstock.com; p. 20 https://commons.wikimedia.org/wiki/File:Mount_St._Helens_1979.jpg; p. 21 https://commons.wikimedia.org/wiki/File:Mount_St._Helens,_one_day_before_the_devastating_eruption.jpg; p. 23 https://commons.wikimedia.org/wiki/File:MSH80_st_helens_eruption_plume_07-22-80.jpg; p. 25 poemnist/Shutterstock.com; p. 26 Francisco Blanco/Shutterstock.com; p. 27 Vara I/Shutterstock.com; p. 29 Tobin Akehurst/Shutterstock.com.

Library of Congress Cataloging-in-Publication Data

Names: Roza, Greg author
Title: The Cascade Range / Greg Roza.
Description: [Buffalo] : PowerKids Press, [2026] | Series: U.S. seismic zones | Includes bibliographical references and index.
Identifiers: LCCN 2025001492 (print) | LCCN 2025001493 (ebook) | ISBN 9781499453300 library binding | ISBN 9781499453294 paperback | ISBN 9781499453317 ebook
Subjects: LCSH: Cascadia Subduction Zone--Juvenile literature | Earthquake zones--Cascade Range--Juvenile
Classification: LCC QE511.46 .R69 2026 (print) | LCC QE511.46 (ebook) | DDC 551.2209795--dc23/eng/20250307
LC record available at https://lccn.loc.gov/2025001492
LC ebook record available at https://lccn.loc.gov/2025001493

Manufactured in the United States of America

CONTENTS

THE CASCADES 4

PLATE TECTONICS 8

THE CASCADIA SUBDUCTION ZONE 10

SEISMIC HISTORY 12

THE 1700 MEGATHRUST 14

THE CIRCUM-PACIFIC BELT 16

THE CASCADE VOLCANIC ARC 18

MOUNT ST. HELENS 20

THE AFTERMATH 24

SEISMIC HOTSPOT 28

GLOSSARY 30

FOR MORE INFORMATION 31

INDEX 32

THE CASCADES

Standing tall along the northwest coast of the Unites States and southwest coast of Canada, the Cascade Range is an area that has undergone earth-shaping seismic activity for millions of years. People travel to this area of the world to see amazing sights, such as snow-capped mountains, giant glaciers, coastal islands, and thick forests.

The Cascade Range is made up of 4,375 named mountains. Earth's most powerful forces helped these mountains grow to towering heights. Many of the peaks were created by volcanic activity. That means that they grew as eruptions sent out **lava**, which hardened to create new landforms. These forces also create thousands of earthquakes in the area every year.

MOUNT RAINIER

The highest peak in the Cascade Range, Mount Rainier, is in the state of Washington. It's an active volcano!

The Cascade Range is about 700 miles (1,127 km) long. It's made up of smaller sections, including the California, Oregon, Washington, and Canadian Cascades. The Cascade Range itself is a smaller part of the main Pacific Mountain System.

Lassen Peak, the southernmost mountain in the range, is a lava dome, which is a steep volcano formed from lava that is very thick. Lassen Peak last erupted in 1917. The northernmost end of the Cascades is near the coast of British Columbia and the top of Vancouver Island in Canada.

The Cascade Range began to form about 37 million years ago, and it will continue to grow for many more years. The forces that build mountains also build volcanoes and create earthquakes. Let's find out how.

EARTH-SHAKING FACT!

Lassen Volcanic National Park is home to all four kinds of volcano. Lassen Peak is a lava dome. Prospect Peak is a shield volcano. Mount Tehama is a composite or stratovolcano. And Cinder Cone is, you guessed it, a cinder cone!

The Pacific Northwest features many locations that are perfect for winter activities and sports. Crystal Mountain in Washington state is only 85 miles (137 km) from Seattle.

TYPES OF VOLCANO

NAME	DESCRIPTION
lava dome	thick lava, slow flow, small with steep sides
shield volcano	thin lava, quick flow, central vent, forms a wide shield shape
stratovolcano	explosive eruptions throw lava and rock, steep sides, main vent and side vents
cinder cone	smallest type, central vent, throws out cinders, ash, and rocks

PLATE TECTONICS

Plate tectonics is the study of Earth's outer layer, known as the lithosphere. This layer is made up of the crust and part of the upper mantle and is about 62 miles (100 km) thick. This rigid layer sits atop the asthenosphere, a layer that is partially molten. This allows sections of the lithosphere, called plates, to drift! These massive plates move ever so slightly. They move away from each other, into each other, and past each other. These actions cause earthquakes and volcanoes.

Pressure builds up between plates that are touching. For a time, often hundreds or thousands of years, two plates remain locked together, and the pressure builds up. Then suddenly, one plate slides against another. This abrupt movement causes an earthquake!

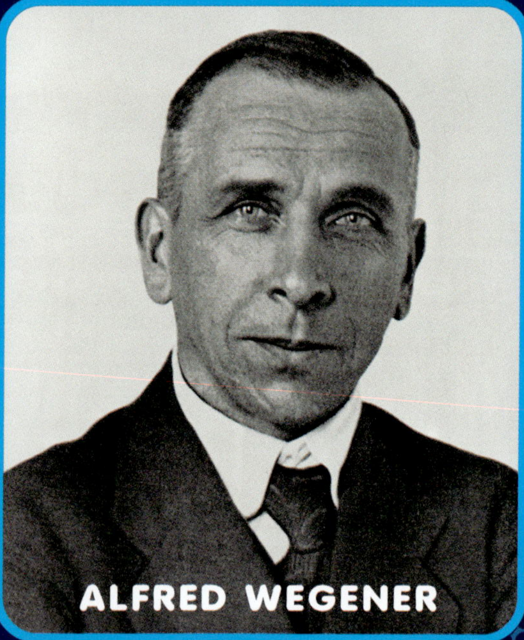

ALFRED WEGENER

Earth's lithosphere is somewhat like an egg with a cracked shell. The different pieces, or plates, have jagged boundaries and fit closely together.

ALFRED WEGENER

Like other scientists of his age, the German **meteorologist** and **geophysicist** Alfred Wegener (1880–1930) noted the similarities in the coastlines of South America and Africa. They look like they once fit together! Wegener theorized that the continents once formed a single landmass he named Pangea. Over a long, long time, the landmasses drifted away from each other, eventually forming the continents we know today. He named this **phenomenon** "continental displacement," which then became "continental drift."

THE CASCADIA SUBDUCTION ZONE

The Cascade Range was formed by a tectonic area known as the Cascadia subduction zone. This describes the type of **fault** that created it. A subduction zone occurs when one or more plates are forced under another plate. For a time, the plates are locked together and pressure grows between them. Once enough pressure builds up, the plates suddenly slide against each other, releasing the pressure, and causing an earthquake. One plate is forced under the other, causing the upper plate to grow in height.

Currently, three small plates are being subducted by the much larger North American plate. The Explorer, Juan de Fuca, and Gorda plates are what's left of the ancient Farallon plate. In time, even these plates will completely disappear beneath the North American plate.

EARTH-SHAKING FACT!

Subduction zones like the one that made the Cascadia Range also produce volcanoes. When sediment and water from the subducting plate push down into the Earth's hot mantle, it melts the rocks, creating magma. The magma gathers and rises to the surface, causing a volcanic eruption.

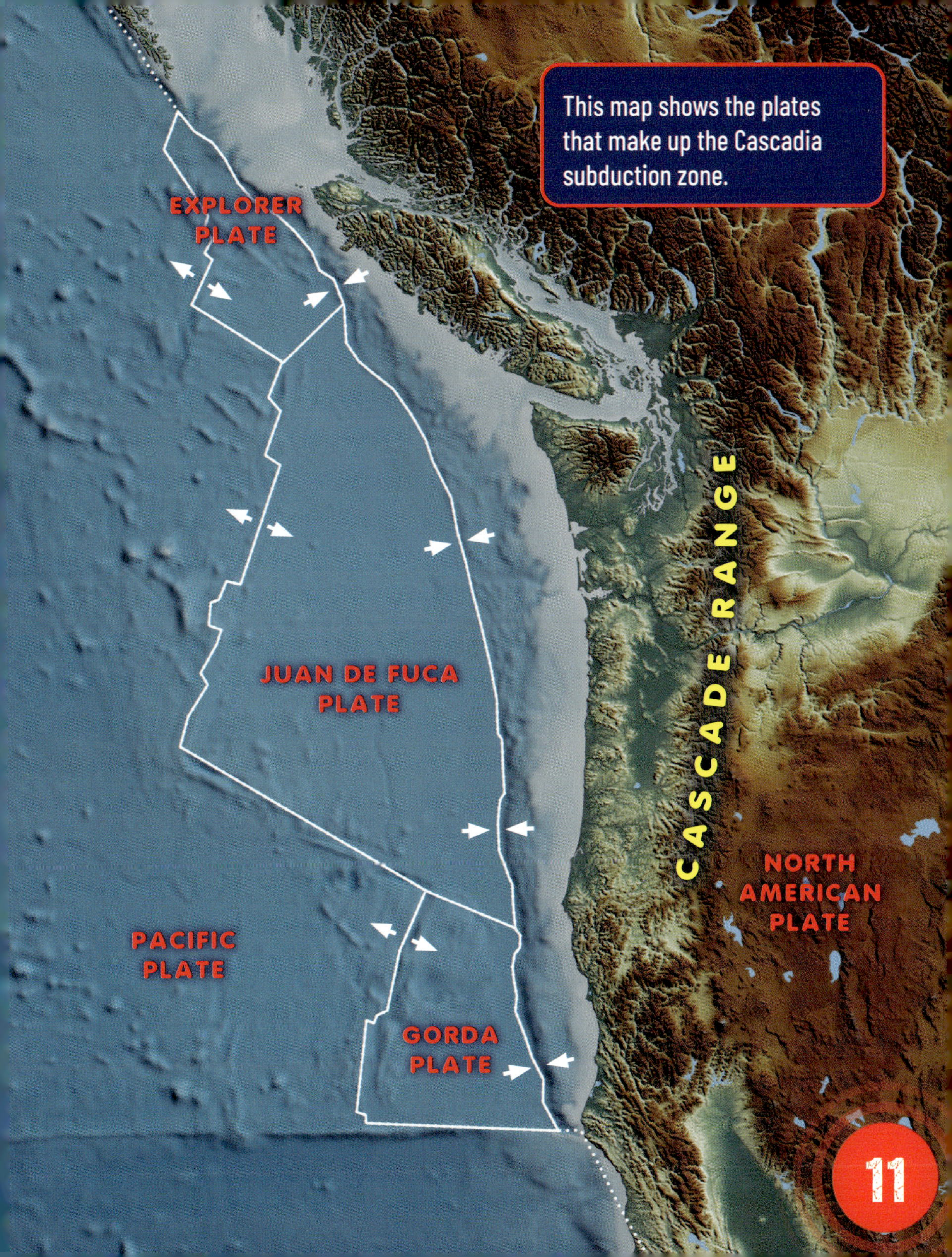

This map shows the plates that make up the Cascadia subduction zone.

SEISMIC HISTORY

Cascadia has long been a seismically active area. For 37 million years, subduction has forced lighter ocean plates beneath the North American plate, resulting in earthquakes, growing mountains, and volcanic peaks. The earliest volcanoes are no longer active; most have been largely eroded away.

The constant subduction process has created several mountain ranges in the North Pacific. In addition to the Cascades, this includes the Olympic Mountains and the Oregon Coast Range. The peaks of the modern Cascade Range formed between 5 and 7 million years ago.

Geologists have discovered evidence of past earthquakes by studying the ocean floor near the Cascadia subduction zone. They have revealed that the area experiences an earthquake with a magnitude greater than 8.0 on the moment magnitude scale approximately once every 500 years.

MOMENT MAGNITUDE SCALE

CATEGORY	MAGNITUDE	DESCRIPTION	HOW COMMON WORLDWIDE?
micro	less than 2.0	Not felt, but recorded by scientific tools	1.4 million a year
minor	2.0 to 2.9	Sometimes felt, rarely cause damage	1.3 million a year
light	3.0 to 3.9	Often felt, rarely causes **significant** damage	130,000 a year
moderate	4.0 to 4.9	Noticeable shaking, significant damage unlikely	13,000 a year
strong	5.0 to 5.9	Can cause significant damage to buildings	1,300 a year
major	6.0 to 6.9	Significant damage in populated areas	100 a year
great	7.0 and higher	Serious damage	10 to 20 times a year

The moment magnitude scale, introduced in the late 1970s, offers a way to measure an earthquake's magnitude, or size. Usually there's only one earthquake per year with a magnitude between 8 and 10, which are called megathrusts. Scientists have never recorded an earthquake higher than 9.5.

THE 1700 MEGATHRUST

The last megathrust in the Cascade Range occurred January 26, 1700. There are no reports of the earthquake from European sources, yet scientists know a great deal about it. Native American groups in the area have stories about terrible shaking, landslides, and **tsunamis** that caused destruction and death. Scientists have discovered the ruins of several Native American communities that **attest** to the stories.

Japan has long kept detailed records of earthquakes and tsunamis. One record tells of a 16-foot (5 m) wall of water striking the east coast around midnight on January 28, 1700. The records refer to the wall of water as an "orphan tsunami," meaning they did not detect the earthquake that caused it. Studies show that the tsunami was caused by the Cascade megathrust and took 10 hours to reach Japan.

The 1700 earthquake had a magnitude of about 9.0. The quake changed the shape of the coastline.

GHOST FOREST NEAR NESKOWIN, OREGON

GHOST FORESTS

Scientists have discovered the stumps of trees near beaches along the northwestern coast. These stumps were once trees that grew near the coast. The subduction, earthquake, and resulting tsunami submerged the trees below water, preserving their stumps. By looking at the rings in the tree stumps, scientists discovered that the last year they grew was 1699.

THE CIRCUM-PACIFIC BELT

The Cascade Range is just a small part of a larger seismic region known as the Circum-Pacific Belt, or more commonly the Ring of Fire. This chain of volcanoes surrounding the Pacific Ocean is about 25,000 miles (40,200 km) long! Both volcanoes and earthquakes are common along the Ring of Fire. In fact, most of Earth's seismic activity occurs in this geological wonder, including 90 percent of all earthquakes and 75 percent of all volcanoes.

There are numerous active volcanoes along the Ring of Fire. Anak Krakatau, shown here, is an active volcano in Indonesia. It last erupted on February 3, 2022.

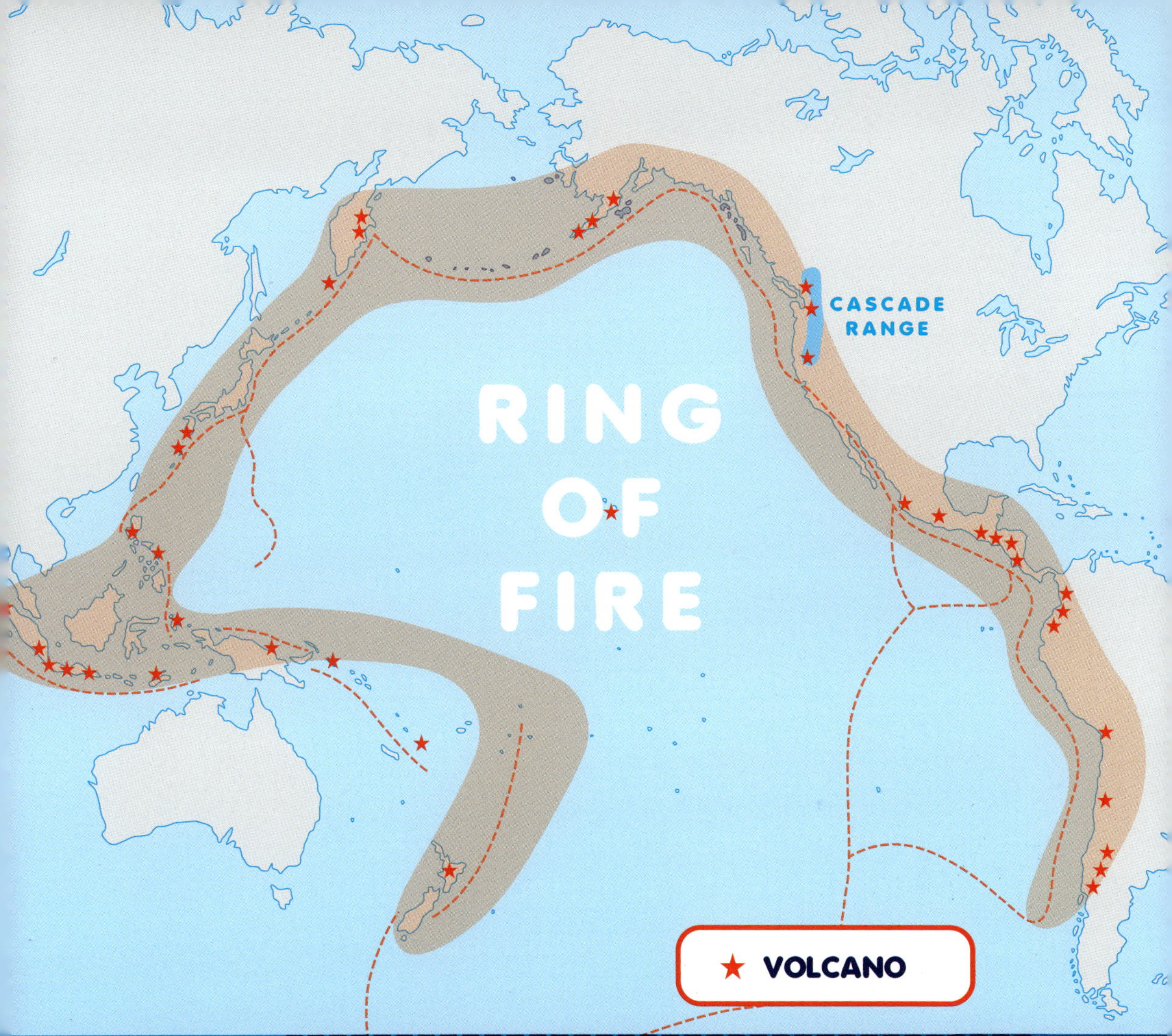

The west coast of North America, including the section known as the Cascade volcanic arc, makes up a large section of the Ring of Fire. This is why so many earthquakes and volcanoes occur along the West Coast. Most of the volcanoes are underwater. Some, such a Mount St. Helens, are in the Cascade Range.

THE CASCADE VOLCANIC ARC

The Cascade volcanic arc has had thousands of volcanoes over the course of its long geological history. Today, it includes about 20 active volcanoes, and more than 4,000 volcanic vents. Seven of the active volcanoes have erupted at least once in the past 200 years. These include (from north to south): Mount Baker, Glacier Peak, Mount Rainier, Mount St. Helens, Mount Hood, Mount Shasta, and Lassen Peak.

Most of the active volcanoes in the Cascade arc are less than 2 million years old. The tallest of them, Mount Rainier, is 14,410 feet (4,392 m) tall. The second tallest is Mount Hood at 11,237 (3,425 m) tall. However, the most **notorious** volcano on the Cascade volcanic arc is Mount St. Helens—the volcano in the Cascades that most recently erupted in 1980.

EARTH-SHAKING FACT!

Crater Lake in Oregon formed in the crater of Mount Mazama. It's the deepest lake in the United States at 1,949 feet (594 m). Wizard Island is the peak formed when Mazama last erupted 4,800 years ago.

A caldera is a wide volcano crater formed when the sides of the volcano break down and fall away. Crater Lake is a caldera filled with water.

MOUNT ST. HELENS

Mount St. Helens in Washington state is a young stratovolcano. Geologists have identified the volcano's oldest ash deposits and dated them to about 40,000 years ago. Over thousands of years, Mount St. Helens rose to a height of 9,677 feet (2,950 m). A series of eruptions ending in 1857 created the Goat Rocks lava dome. The peak remained this way until the early 1980s when seismic activity around it increased.

Mount St. Helens, shown here the day before it erupted in 1980, had 11 named glaciers and 2 unnamed glaciers. The eruption destroyed about 70 percent of the glaciers. However, a new glacier has formed in the crater. In 2006, it was named Crater Glacier.

Beginning on March 16, 1980, the area around Mount St. Helens experienced thousands of earthquakes. Hundreds of steam explosions began to appear around the mountain. On the north side of the mountain, a bulge of land grew over 260 feet (79 m) high! After a series of moderate to strong earthquakes, it was clear something big was about to happen.

At 8:32 a.m., May 18, 1980, an earthquake occurred about a mile beneath Mount St. Helens. Seconds later, the top 1,300 feet (396 m) of the mountain exploded, leaving a mile-wide crater. It created the largest landslide in modern times. One estimate says that the debris in the landslide could fill 1 million Olympic-sized swimming pools!

The eruption created a massive cloud of ash that reached 12 miles (19 km) into the sky. **Pyroclastic flows** rained smoke, rock, and volcanic gasses down on the surrounding area. All plants and trees within in a 6-mile (9.7 km) area were completely destroyed.

The eruption killed 57 people. These included loggers, campers, and reporters. It also led to the death of David A. Johnston, a brave volcanologist who worked for the U.S. Geological Survey (USGS).

DAVID A. JOHNSTON

Johnston was observing Mount St. Helens from a place 6 miles (9.7 km) away. The sudden eruption, combined with the speed at which the pyroclastic flow was moving, surprised Johnston and others who were nearby. Johnston described the quickly approaching flows to coworkers through radio contact. His final words were: "It's going to hit me, too." Today, several areas in Mount St. Helens National Monument are named for Johnston.

Most of the animals on and around Mount St. Helens died in the eruption. This included birds, fish, small mammals, elk, deer, bears, and more. However, many burrowing animals, including gophers, stayed alive by hiding underground.

THE AFTERMATH

The ash released from Mount St. Helens circled the planet several times before falling. Much of the ash piled up around the mountain. It took more than two months and millions of dollars to clean it up. Much of the debris was dumped in quarries and landfills.

For the next five years, Mount St. Helens remained active. More eruptions occurred and lava flows in the crater made new lava domes. Some of those lava domes were destroyed by later eruptions. By 1986, a new lava dome had grown to 820 feet (250 m).

Although seismic activity has lessened since then, Mount St. Helens is still active. Smaller explosions occurred between 1989 and 1991, in 1995, and 1998. In the early 2000s, small explosions and lava flows created new lava domes. The last continuous volcanic activity occurred in 2008 when steam erupted from a lava dome.

Since 2018, there have been at least 40 earthquakes in the area.

Mount St. Helens National Volcanic Monument, founded in 1982, offers tourists the opportunity to view the amazing landscape and learn about the events that occurred in May 1980. The Johnston Ridge **Observatory** is on the mountain's west side. It allows tourists to view the U-shaped crater, the lava dome, and the effects landslides had on the landscape.

JOHNSTON RIDGE OBSERVATORY

This is the view tourists see when they visit Johnston Ridge Observatory, named after David A. Johnston.

This area is also very important to scientists and others who study volcanoes and active seismic zones. Mount St. Helens is still the most seismically active location in the Cascade Range. The network of research stations on and around the mountain were established in March of 1980 in response to the increased seismic activity. The scientific instruments that record seismic activity in the area have been updated over the years.

SEISMIC HOTSPOT

Seismologists believe that the Cascade subduction zone will create another megathrust earthquake in the next 150 years. This could bring widespread destruction to major cities along North America's west coast, from both the quake and the resulting tsunami. Coastal cities such as Vancouver, British Columbia; Seattle, Washington; Tacoma, Washington; and Portland, Oregon, might experience destructive seismic activity. In addition, scientists believe that Mount St. Helens, as well as other volcanoes in the Cascade Range, is likely to have another large eruption in the coming years.

Volcanologists are constantly studying the Cascade subduction zone and the landforms it has created over the course of millions of years. They hope to be able to have enough warning before a major seismic event occurs so they can warn people of the approaching danger.

EARTH-SHAKING FACT!

In 2019, the USGS set up the National Volcano Early Warning System (NVEWS), which monitors the country's most active volcanoes. The goal is to provide people plenty of warning before volcanic activity occurs.

The Cascades offer many breathtaking sights for tourists, hikers, skiiers, hunters, and more. How long will it be until earthquakes and volcanoes reshape the land?

GLOSSARY

attest: To confirm that something is true.

cinder: A bit of cooled lava.

fault: A break or crack in Earth's crust.

geophysicist: A scientist that deals with the physical processes occurring in or near the earth

lava: Molten (melted) rock that reaches Earth's surface.

magma: Molten (melted) rock beneath Earth's surface.

meteorologist: A scientist who studies the Earth's atmosphere, weather, and climate.

notorious: Widely and unfavorably known.

observatory: A building or other structure that provides a wide view of the surroundings.

phenomenon: A fact or event that is observed.

pyroclastic flow: An extremely hot mixture of rock, dust, and gas from a volcano that moves at high speed away from the volcano when it erupts.

significant: A noticeably large amount.

tsunami: A great sea wave created by an earthquake or volcanic eruption under the sea.

FOR MORE INFORMATION

BOOKS

Adamson, Thomas K. *The Eruption of Mount St. Helens*. Minnetonka, MN: Bellwether Media, 2021.

Izzy, Howell. *Earthquakes*. Buffalo, NY: Enslow Publishing, 2024.

Izzy, Howell. *Volcanoes*. Buffalo, NY: Enslow Publishing, 2024.

WEBSITES

Kiddle: Cascadia Subduction Zone Facts for Kids
kids.kiddle.co/Cascadia_subduction_zone
Learn more fascinating facts about the Cascadia subduction zone here.

The Legacy of David A. Johnston
www.usgs.gov/observatories/cascades-volcano-observatory/legacy-david-johnston
Read about the dedication and bravery of David A. Johnston and his contributions to science.

Publisher's note to educators and parents: Our editors have carefully reviewed these websites to ensure that they are suitable for students. Many websites change frequently, however, and we cannot guarantee that a site's future contents will continue to meet our high standards of quality and educational value. Be advised that students should be closely supervised whenever they access the internet.

INDEX

A

asthenosphere, 8

B

British Columbia, 6, 28

C

Crater Lake, 18, 19

E

Explorer plate, 10

F

fault, 10

G

ghost forests, 15
glaciers, 4, 21
Gorda plate, 10

J

Johnston, David A., 22, 27
Johnston Ridge Observatory, 26, 27
Juan de Fuca plate, 10

L

landslides, 14, 22, 26
Lassen Peak, 6, 18
lithosphere, 8, 9

M

megathrusts, 13, 14, 28
moment magnitude scale, 12, 13
Mount Hood, 18
Mount Rainier, 4, 5, 18
Mount St. Helens, 17, 18, 20, 21, 22, 23, 24, 26, 27, 28

N

North American plate, 10, 12

P

Pacific Mountain System, 6
plate tectonics, 8

R

Ring of Fire, 16, 17

S

seismologist, 28
1700 earthquake, 14, 15
subduction zone, 10, 11, 12, 28

T

tsunamis, 14, 15, 28
types of volcanoes, 6, 7

U

U.S. Geological Survey (USGS), 22

V

Vancouver Island, 6
volcanologist, 22, 28

W

Wegener, Alfred, 9